Neill Hamilton

CCEA GCSE

FURTHER MATHS REVISION BOOKLET

PURE MATHS 1

COLOURPOINT
EDUCATIONAL

Contents

Revision Exercise 1

1. Matrices A, B and C are defined as:

$A = \begin{pmatrix} -2 & 5 \\ 3 & -4 \end{pmatrix}$ $B = \begin{pmatrix} 2 & -1 \\ 5 & -3 \end{pmatrix}$ $C = \begin{pmatrix} -6 & 4 \\ 3 & -2 \end{pmatrix}$

Work out:

(a) A – B

Answer ______________________________ [1]

(b) 3A + C

Answer ______________________________ [1]

(c) AB

Answer ______________________________ [2]

2. Matrices P and Q are defined as:

$P = \begin{pmatrix} 5 & -2 \\ -1 & -3 \end{pmatrix}$ $Q = \begin{pmatrix} -2 & -4 \\ 3 & -2 \end{pmatrix}$

Solve:

(a) $X = P - Q$

Answer ________________________ [1]

(b) $3P + Y = Q$

Answer ________________________ [2]

(c) $Q - 2Z = 4P$

Answer ________________________ [2]

3. Differentiate:

(a) $5x^2 - x + \frac{1}{x^2}$

Answer ________________________ [3]

(b) $\frac{3x^8}{4} + 2x - \frac{3}{x^4}$

Answer ________________________ [3]

(c) $\frac{5x^2}{6} + \frac{4}{9x^3}$

Answer ________________________ [2]

4. Work out $\int(6x - 5)^2\,dx$

Answer ______________________ [4]

5. The equation of a curve is $y = 2x^3 - 5x^2 + x + 7$
Find:
(a) the gradient at the point (2, 5)

Answer ______________________ [3]

(b) the equation of the tangent to the curve at the point (2, 5)

Answer ______________________ [2]

(c) the equation of the normal to the curve at the point (2, 5).

Answer ______________________ [3]

6. The equation of a curve is $y = x^2 + 3x - 4$

(a) Find where the curve crosses the x-axis.

Answer ______________________ [3]

(b) Find where the curve crosses the y-axis.

Answer ______________________ [1]

(c) Find the coordinates of the turning point of the curve.

Answer ______________________ [4]

(d) Use calculus to show whether the turning point is a maximum or a minimum.

Answer ______________________ [2]

(e) Sketch the curve.

[2]

7. Find the area under the curve $y = 15 + 2x - x^2$ between $x = -1$ and $x = 2$

Answer ______________________ [4]

8. David took out an investment in 2020.

The cash–in value, £C, of this investment can be modelled by $C = 800 + 17t - 2t^2$ where t is the time in years since 2020.

Work out the maximum cash–in value of this investment, justifying your answer.

Answer ____

____ [5]

Total for revision exercise **[50]**

Revision Exercise 2

1. Matrices A, B, C and D are defined as:

$A = \begin{pmatrix} 6 \\ 3 \end{pmatrix}$ $B = \begin{pmatrix} -2 & 7 \end{pmatrix}$ $C = \begin{pmatrix} 1 & -2 \\ 4 & -3 \end{pmatrix}$ $D = \begin{pmatrix} -3 & 1 & -4 \\ 2 & 5 & -2 \end{pmatrix}$

Say whether the following products exist. If they do, work them out.

(a) AB

Answer ______________________ [1]

(b) BA

Answer ______________________ [1]

(c) AC

Answer ______________________ [1]

(d) CA

Answer ______________________ [1]

(e) CB

Answer ______________________ [1]

(f) DC

Answer ______________________ [1]

(g) CD

Answer ______________________ [1]

(h) B^2

Answer ______________________ [1]

(i) C^2

Answer ______________________ [1]

2. Matrices P and Q are defined as:

$P = \begin{pmatrix} 3 & -5 \\ 2 & -4 \end{pmatrix} \qquad Q = \begin{pmatrix} -5 & 1 \\ 6 & -2 \end{pmatrix}$

Solve:

(a) PX = Q

Answer ______________________ [2]

(b) QX = P

Answer ______________________ [2]

3. Find $\frac{d^2y}{dx^2}$ if

(a) $y = 4x^2 - \frac{2}{5x^2}$

Answer ______________________ [3]

(b) $y = \frac{5x^2}{6} + \frac{3}{2x^5}$

Answer ______________________ [3]

4. $\int\left(\frac{7}{3}x^5 - \frac{2}{7x^2}\right)dx$

Answer ______________________ [3]

5. **(a)** Find the coordinates of the point where the gradient of the tangent to the curve
$y = 2x^2 + 3x$
is 5.

Answer ______________________ [3]

(b) Find the coordinates of the points where the tangents to the curve
$y = 4x^3 + 33x^2 - 36x + 2$
are horizontal.

Answer ______________________ [4]

6. The equation of a curve is $y = (5 - x)(4 + x)$.
(a) Find where the curve crosses the x-axis.

Answer ______________________ [2]

(b) Find where the curve crosses the y-axis.

Answer ______________________ [1]

(c) Find the coordinates of the turning point of the curve.

Answer ______________________ [4]

(d) Use calculus to show whether the turning point is a maximum or a minimum.

Answer ______________________________ [2]

(e) Sketch the curve.

[2]

7. **(a)** Find where the curve $y = 2x^2 - x - 15$ crosses the x-axis.

Answer ______________________________ [2]

(b) Hence find the area between the curve $y = 2x^2 - x - 15$ and the x-axis.

Answer ______________________________ [3]

8. The length and breadth of a rectangle are $(4x + 3)$ cm and $(3 - x)$ cm.
Work out the maximum possible area, justifying your answer.

Answer ____

____ [5]

Total for revision exercise **[50]**

Revision Exercise 3

1. Work out:

(a) $\begin{pmatrix} x \\ 3x \end{pmatrix} + \begin{pmatrix} -4x \\ 2 \end{pmatrix}$

Answer ______________________ [1]

(b) $\begin{pmatrix} 3 \\ y-2 \end{pmatrix} - \begin{pmatrix} 2y \\ y \end{pmatrix}$

Answer ______________________ [1]

(c) $\begin{pmatrix} v-4 \\ 3v \end{pmatrix} - \begin{pmatrix} 2v \\ 2 \end{pmatrix}$

Answer ______________________ [1]

2. Use matrices to solve $3x + 4y = -4$ and $2x - 3y = -14$

Answer ______________________ [5]

3. Find $\frac{dy}{dx}$ if $y = ax^3 - \frac{b}{x}$

Answer ______________________ [2]

4. The gradient function of a curve is given by $3x^2 + 4x - \frac{5}{x^2}$ at any point (x, y).
The curve passes through (1, 2).
Find the equation of the curve.

Answer ______________________ [7]

5. (a) Find the equation of the tangent to the curve $y = \frac{5}{16}x^2 - 2x + 8$ at the point whose x-coordinate is 4.

Answer ______________________ [6]

(b) Hence find where this tangent crosses the
(i) y-axis

Answer ______________________ [1]

(ii) x-axis

Answer ______________________ [1]

6. The equation of a curve is $y = x(2x^2 + 6x + 5)$

(a) Show that the curve crosses the x-axis at one point only.

Answer ______________________ [4]

(b) Find the coordinates of the turning points of the curve, giving your answers correct to 2 decimal places.

Answer ______________________ [5]

(c) Use calculus to show whether each turning point is a maximum or a minimum.

Answer ______________________ [2]

(d) Sketch the curve.

[2]

7. The equation of a curve is $y = 6x^2 - 7x$
Find the value of k for which the area between $y = 6x^2 - 7x$, the x-axis, $x = 0$ and $x = \frac{k}{2}$ is one third of the area between $y = 6x^2 - 7x$, the x-axis, $x = 0$ and $x = k$.

Answer ______________________ [7]

8. Find the value of x for which the product of $(3x - 2)$ and $(2x + 5)$ will be a minimum.
Justify your answer.

Answer ____

____ [5]

Total for revision exercise **[50]**

Revision Exercise 4

1. Matrices A,B and C are defined as:

$$A = \begin{pmatrix} -2 & 5 \\ 3 & -4 \end{pmatrix} \qquad B = \begin{pmatrix} 2 & -1 \\ 5 & -3 \end{pmatrix} \qquad C = \begin{pmatrix} -6 & 4 \\ 3 & -2 \end{pmatrix}$$

(a) Work out:

(i) C^2

Answer ______________________ [1]

(ii) $B(C - 2A)$

Answer ______________________ [2]

(b) Which one of the matrices A, B or C has no inverse?

Answer ______________________ [1]

2. Use matrices to prove that the lines $6x - 4y = 12$ and $3x - 2y = 5$ do not cross.

[4]

3. Find $\frac{d^2y}{dx^2}$ if

$y = kx^5 + \frac{m}{x^2}$

Answer ____________________ [4]

4. Work out $\int_1^2 (4x^3 - \frac{3}{4x^3})dx$

Answer ____________________ [4]

5. **(a)** Find the equation of the tangent to the curve $y = 8 - 3x + x^2$ at the point $(-5, 48)$.

Answer ____________________ [4]

(b) Hence find where this tangent crosses the line $y = 7x + 3$

Answer ____________________ [3]

6. The equation of a curve is $y = 6x^3 + 3x^2 - 30x$

(a) Find where the curve crosses the x-axis.

Answer ______________________ [4]

(b) Find the coordinates of the turning points of the curve, giving your answers correct to 2 decimal places.

Answer ______________________ [5]

(c) Use calculus to show whether each turning point is a maximum or a minimum.

Answer ______________________ [2]

(d) Sketch the curve.

[2]

7. The equation of a curve is $y = 2x^2 + 7x - 4$

   (a) Find where the curve crosses the x-axis.

   Answer ______________________ [2]

   (b) Find the area between the curve, the positive x-axis and the negative y-axis.

   Answer ______________________ [3]

8. The length and breadth of a rectangle are $(3x - 2)$ cm and $(2y + 1)$ cm. Its perimeter is 20 cm.

   (a) Write y in terms of x.

   Answer ______________________ [2]

   (b) Work out the area of the rectangle in terms of x.

   Answer ______________________ [2]

(c) Hence find the maximum area of the rectangle, showing that this area is a maximum.

Answer ______________________________ [5]

Total for revision exercise **[50]**

Revision Exercise 5

1. Matrices A, B and C are defined as:

$A = \begin{pmatrix} -2 & 5 \\ 1 & -3 \end{pmatrix}$ $B = \begin{pmatrix} -1 & 4 \\ 2 & -1 \end{pmatrix}$ $C = \begin{pmatrix} 10 & -14 \\ 5 & -2 \end{pmatrix}$

Find the value of AB – C

Answer ______________________ [3]

2. Matrices P and Q are defined as:

$P = \begin{pmatrix} 3 & -5 \\ 2 & -4 \end{pmatrix}$ $Q = \begin{pmatrix} -5 & 1 \\ 6 & -2 \end{pmatrix}$

Solve:

(a) PX = Q + 2P

Answer ______________________ [4]

(b) QX = 3P – PX

Answer ______________________ [4]

3. $y = 4x^3 - \frac{5a}{x} - 3$

Find the value of a for which $\frac{dy}{dx} = 7$ when $x = -2$

Answer ______________________ [4]

4. Work out $\int_1^3 (x^2 + \frac{1}{5x^4})dx$

Answer ______________________ [4]

5. Find the coordinates of the other point where the normal to the curve $y = 2x^2 - 5x - 3$ at $(2, -5)$ crosses the curve.

Answer ______________________ [5]

6. The equation of a curve is $y = (x - 2)(x + 4)(2x - 1)$
   (a) Find where the curve crosses the x-axis.

Answer ______________________ [3]

   (b) Find the coordinates of the turning points of the curve, giving your answers correct to 2 decimal places.

Answer ______________________ [5]

   (c) Use calculus to show whether each turning point is a maximum or a minimum.

Answer ______________________ [2]

   (d) Sketch the curve.

[2]

7. The area under the curve $y = ax^3 - 2x + 6$ between $x = 0$ and $x = 2$ is 24.
   Find the value of a.

Answer ______________________ [5]

8. An object travels at an average speed of x m/s for $(x + 1)$ sec and then at an average speed of $(2 - x)$ m/s for $(6 - x)$ sec.

(a) Find the total distance, D m, travelled in terms of x.

Answer ______________________ [2]

(b) Find the value of x for which this distance is a minimum, proving that this value gives a minimum.

Answer ______________________ [4]

(c) Find the minimum distance travelled.

Answer ______________________ [1]

Total for revision exercise **[50]**

Answers

Revision Exercise 1

1. (a) $\begin{pmatrix} -4 & 6 \\ -2 & -1 \end{pmatrix}$ [1] (b) $\begin{pmatrix} -12 & 19 \\ 12 & -14 \end{pmatrix}$ [1]

(c) $\begin{pmatrix} 21 & -13 \\ -14 & 9 \end{pmatrix}$ [2] or [1] for 2 or 3 correct values.

2. (a) $\begin{pmatrix} 7 & 2 \\ -4 & -1 \end{pmatrix}$ [1]

(b) Y = Q − 3P so $\begin{pmatrix} -17 & 2 \\ 6 & 7 \end{pmatrix}$ [M1W1]

(c) 2Z = Q − 4P so 2Z = $\begin{pmatrix} -22 & 4 \\ 7 & 10 \end{pmatrix}$ so Z = $\begin{pmatrix} -11 & 2 \\ 3.5 & 5 \end{pmatrix}$ [M1W1]

3. (a) $10x - 1 - \frac{2}{x^3}$ [3] (b) $6x^7 + 2 + \frac{12}{x^5}$ [3]

(c) $\frac{5x}{3} - \frac{4}{3x^4}$ [2]

4. $\int(36x^2 - 60x + 25)\,dx = 12x^3 - 30x^2 + 25x + c$ [1] for correct expansion and [3] for correct answers

5. $\frac{dy}{dx} = 6x^2 - 10x + 1$

(a) $x = 2$ so $m = 24 - 20 + 1 = 5$ [M1W2]

(b) $y = 5x + c$, so $5 = 10 + c$, so $c = -5$

Answer: $y = 5x - 5$ [M1W1] (c) $y = -\frac{1}{5}x + c$, so $5 = -\frac{2}{5} + c$, so $c = 5\frac{2}{5}$

Answer: $y = -\frac{1}{5}x + 5\frac{2}{5}$ [M1W2]

6. (a) $x^2 + 3x - 4 = 0$, $(x + 4)(x - 1) = 0$, $x = -4$ or 1

Answer: (−4, 0) and (1, 0) [M1W2]

(b) $x = 0$ so $y = 0 + 0 - 4 = -4$, Answer: (0, −4) [1]

(c) $\frac{dy}{dx} = 2x + 3 = 0$, so $x = -\frac{3}{2}$, so $y = -6\frac{1}{4}$

Answer: $(-\frac{3}{2}, -6\frac{1}{4})$ [M3W1]

(d) $\frac{d^2y}{dx^2} = 2$ so minimum turning point [M1W1]

(e)

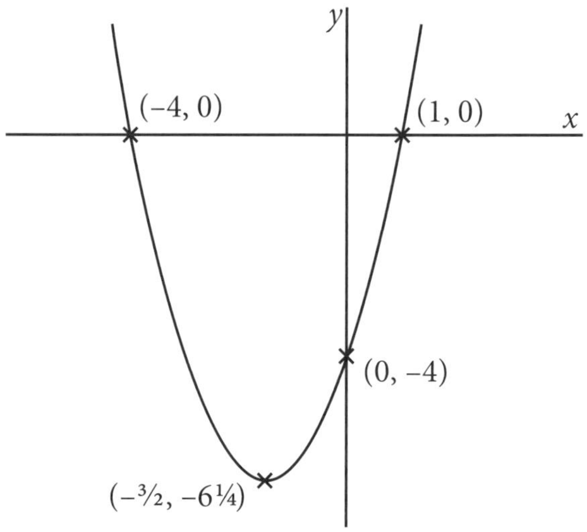

[1] for points, [1] for curve

7. $\int_{-1}^{2}(15 + 2x - x^2)dx = 45$ [M3W1]

8. $\frac{dC}{dt} = 17 - 4t = 0$ [M1W1] so $t = \frac{17}{4}$ [1]

Answer: £836.12 or £836.13 [1] because $\frac{d^2C}{dt^2} = -4$ so a maximum turning point. [1]

Revision Exercise 2

1. (a) Yes $\begin{pmatrix} -12 & 42 \\ -6 & 21 \end{pmatrix}$ [1] (b) Yes $\begin{pmatrix} 9 \end{pmatrix}$ [1] (c) No [1]

(d) Yes $\begin{pmatrix} 0 \\ 15 \end{pmatrix}$ [1] (e) No [1] (f) No [1]

(g) Yes $\begin{pmatrix} -7 & -9 & 0 \\ -18 & -11 & -10 \end{pmatrix}$ [1] (h) No [1]

(i) Yes $\begin{pmatrix} -7 & 4 \\ -8 & 1 \end{pmatrix}$ [1]

2. (a) $\begin{pmatrix} -25 & 7 \\ -14 & 4 \end{pmatrix}$ [2] or [1] for 2 or 3 correct

(b) $\begin{pmatrix} -2 & 3.5 \\ -7 & 12.5 \end{pmatrix}$ [2] or [1] for 2 or 3 correct

3. (a) $y = 8 - \frac{12}{5x^4}$ [M1W2] (b) $\frac{5}{3} + \frac{45}{x^7}$ [M1W2]

4. $\frac{7}{18}x^6 + \frac{2}{7x} + c$ [3]

5. (a) $\frac{dy}{dx} = 4x + 3 = 5$ so $x = \frac{1}{2}$, Answer: $(\frac{1}{2}, 2)$ [M1W2]

(b) $\frac{dy}{dx} = 12x^2 + 66x - 36 = 0$

so $2x^2 + 11x - 6 = 0$, so $(2x - 1)(x + 6) = 0$

Answer: $(\frac{1}{2}, -7\frac{1}{4})$ and (−6, 542) [M2W2]

6. (a) $(-4, 0)$ and $(5, 0)$ [2] (b) $(0, 20)$ [1]
(c) $(\frac{1}{2}, 20\frac{1}{4})$ [M3W1]
(d) $\frac{d^2y}{dx^2} = -2$ so maximum turning point [M1W1]
(e)

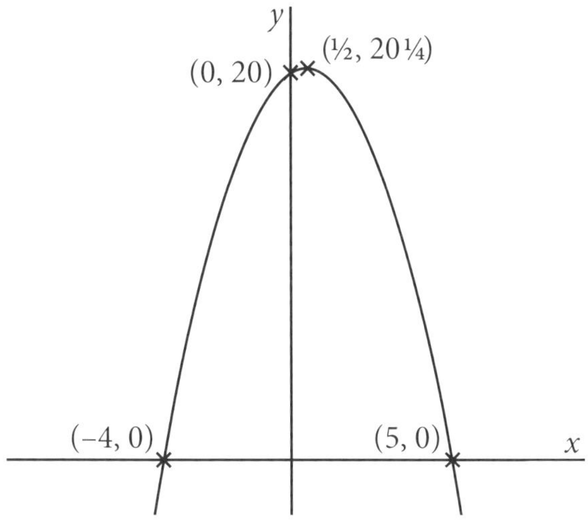

[1] for points, [1] for curve

7. (a) $(-\frac{5}{2}, 0)$ and $(3, 0)$ [M1W1]
(b) $\int_{-5/2}^{3}(2x^2 - x - 15)dx = -55\frac{11}{24}$
so area $= 55\frac{11}{24}$ [M2W1]

8. Area $A = (4x + 3)(3 - x) = 9x - 4x^2 + 9$
$\frac{dA}{dx} = 9 - 8x = 0$ [M1W1], so $x = \frac{9}{8}$ [1]
Answer: $14\frac{1}{16}$ cm^2 [1] because $\frac{d^2A}{dx^2} = -8$ so a maximum turning point. [1]

Revision Exercise 3

1. (a) $\begin{pmatrix} -3x \\ 3x+2 \end{pmatrix}$ [1] (b) $\begin{pmatrix} 3-2y \\ -2 \end{pmatrix}$ [1] (c) $\begin{pmatrix} -v-4 \\ 3v-2 \end{pmatrix}$ [1]

2. $\begin{pmatrix} 3 & 4 \\ 2 & -3 \end{pmatrix}\begin{pmatrix} x \\ y \end{pmatrix} = \begin{pmatrix} -4 \\ -14 \end{pmatrix}$ det $= -9 - 8 = -17$
so $x = -\frac{1}{17}\begin{pmatrix} -3 & -4 \\ -2 & 3 \end{pmatrix}\begin{pmatrix} -4 \\ -14 \end{pmatrix} = \begin{pmatrix} -4 \\ 2 \end{pmatrix}$ [M2W3]

3. $3ax^2 + \frac{b}{x^2}$ [2]

4. $\frac{dy}{dx} = 3x^2 + 4x - 5x^{-2} = 0$
$y = x^3 + 2x^2 + 5x^{-1} + c$, $2 = 1 + 2 + 5 + c$, $c = -6$
Answer: $y = x^3 + 2x^2 + 5x^{-1} + 6$ [M1W6]

5. (a) $y = \frac{1}{2}x + 3$ [M1W5] (b) (i) $x = 0$ so $y = 0 + 3$, Answer $(0, 3)$ [1] (ii) $y = 0$ so $\frac{1}{2}x + 3 = 0$ so $x = -6$, Answer $(-6, 0)$ [1]

6. (a) $2x^3 + 6x^2 + 5x = 0$, $x(2x^2 + 6x + 5) = 0$,
so $x = 0$ or $x = \frac{-6 \pm \sqrt{36 - 40}}{4}$ but $\sqrt{-4}$ does not exist so only crosses x–axis at $(0, 0)$. [M1W3]
(b) $(-0.59, -1.27)$ and $(-1.41, -0.73)$ [M1W4]
(c) When $x = -0.59$, $\frac{d^2y}{dx^2} = 4.92$, so $(-0.59, -1.27)$ is a minimum turning point. When $x = -1.41$, $\frac{d^2y}{dx^2} = -4.92$, so $(-1.41, -0.73)$ is a maximum turning point. [M1W1]
(d)

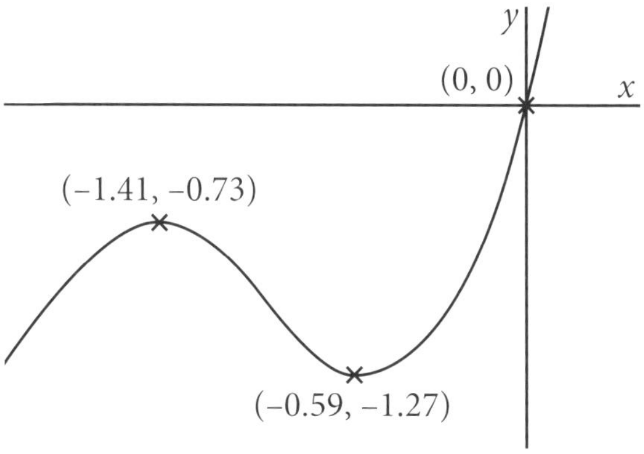

[1] for points, [1] for curve

7. $\int_0^{k/2}(6x^2 - 7x)dx = \frac{1}{4}k^3 - \frac{7}{8}k^2$
$\int_0^{k}(6x^2 - 7x)dx = 2k^3 - \frac{7}{2}k^2$
$\frac{1}{4}k^3 - \frac{7}{8}k^2 = \frac{1}{3}(2k^3 - \frac{7}{2}k^2)$
$10k^3 - 7k^2 = 0$
$k^2(10k - 7) = 0$
Answer: $k = \frac{7}{10}$ [M1W6]

8. Product $P = (3x - 2)(2x+5) = 6x^2 + 11x - 10$
$\frac{dP}{dx} = 12x + 11$ so $x = -\frac{11}{12}$
$\frac{d^2P}{dx^2} = 12$, so a minimum turning point. [M1W4]

Revision Exercise 4

1. (a) (i) $\begin{pmatrix} 48 & -32 \\ -24 & 16 \end{pmatrix}$ [1] (ii) $\begin{pmatrix} -1 & -18 \\ -1 & -48 \end{pmatrix}$ [2]
(b) C because det C = 12 – 12 = 0 [1]

2. Determinant of $\begin{pmatrix} 6 & -4 \\ 3 & -2 \end{pmatrix} = -12 - (-12) = 0$, so no solution, so they don't cross. [M2W2]

3. $20kx^3 + \frac{6m}{x^4}$ [M2W2]

4. $\int_1^2(4x^3 - \frac{3}{3x^3})dx = \left[x^4 + \frac{3}{8x^2}\right]_1^2 = 16\frac{3}{32} - 1\frac{3}{8} = 14\frac{23}{32}$
[M1W3]

5. (a) $y = -13x - 17$ [M1W3]
(b) $-13x - 17 = 7x + 3$, so $x = -1$.
$y = 13 - 17 = -4$, Answer: $(-1, -4)$ [M1W2]

6. (a) $6x^3 + 3x^2 - 30x = 0$, $3x(2x^2 + x - 10) = 0$,
$3x(2x + 5)(x - 2) = 0$, so $x = 0$ or $-\frac{5}{2}$ or 2
Answer: $(0, 0)$, $(-\frac{5}{2}, 0)$, $(2, 0)$ [M1W3]

(b) $\frac{dy}{dx} = 18x^2 + 6x - 30 = 0$

Answer: (1.14, −21.41) and (−1.47, 31.52) [M1W4]

(c) When $x = 1.14$, $\frac{d^2y}{dx^2} = 47.04$ so (1.14, −21.41) is a minimum turning point.

When $x = -1.47$, $\frac{d^2y}{dx^2} = -46.92$ turning point, so (−1.47, 31.52) is a maximum turning point. [M1W1]

(d)

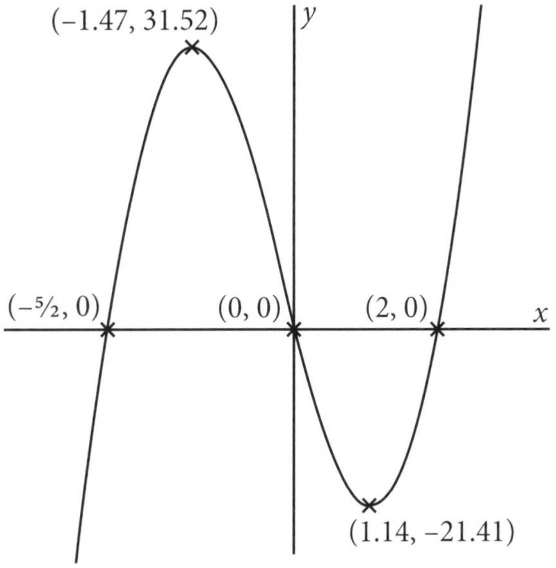

[1] for points, [1] for curve

7. (a) ($\frac{1}{2}$, 0),(−4, 0) [M1W1] (b) 1 $\frac{1}{24}$ [M1W2]
8. (a) $2(3x - 2) + 2(2y + 1) = 20$, $6x + 4y - 2 = 20$, $4y = 22 - 6x$, $y = \frac{11}{2} - \frac{3}{2}x$ [M1W1]

(b) $A = (3x - 2)(12 - 3x) = 42x - 9x^2 - 24$

$\frac{dA}{dx} = 42 - 18x$, $\frac{d^2A}{dx^2} = -18$, so a maximum.

$42 - 18x = 0$ so $x = \frac{7}{3}$ [M1W1]

(c) $A = 42x - 9x^2 - 24 = (42 \times \frac{7}{3}) - (9 \times \frac{49}{9}) - 24 = 25$ cm^2 [M1W4]

Revision Exercise 5

1. $\begin{pmatrix} 12 & -13 \\ -7 & 7 \end{pmatrix} - \begin{pmatrix} 10 & -14 \\ 5 & -2 \end{pmatrix} = \begin{pmatrix} 2 & 1 \\ -12 & 9 \end{pmatrix}$ [M2W]

2. (a) $-\frac{1}{2}\begin{pmatrix} -4 & 5 \\ -2 & 3 \end{pmatrix}\begin{pmatrix} 1 & -9 \\ 10 & -10 \end{pmatrix} = \begin{pmatrix} -23 & 7 \\ -14 & 6 \end{pmatrix}$

[M2W2]

(b) (Q + P)X = 3P so $\begin{pmatrix} -2 & -4 \\ 8 & -6 \end{pmatrix} X = \begin{pmatrix} 9 & -15 \\ 6 & -12 \end{pmatrix}$

so $X = \frac{1}{44}\begin{pmatrix} -30 & 42 \\ -84 & 144 \end{pmatrix}$ [M2W2]

3. $\frac{dy}{dx} = 12x^2 + \frac{5a}{x^2}$, so $48 + \frac{5a}{4} = 7$, giving $a = -32.8$

[M1W2]

4. $8\frac{296}{405}$ [M1W3]

5. $\frac{dy}{dx} = 4x - 5$, so gradient of tangent is 3,

so gradient of normal is $-\frac{1}{3}$

so equation of normal is $y = -\frac{1}{3}x - \frac{13}{3}$

so $2x^2 - 5x - 3 = -\frac{1}{3}x - \frac{13}{3}$

so $6x^2 - 15x - 9 = -x - 13$

so $6x^2 - 14x + 4 = 0$

so $3x^2 - 7x + 2 = 0$

so $(3x - 1)(x - 2)$ giving $x = \frac{1}{3}$

Answer: ($\frac{1}{3}$, $-\frac{40}{9}$) [M2W5]

6. (a) (−4, 0), ($\frac{1}{2}$, 0), (2,0) [3]

(b) (1.30, −5.94), (−2.30, 40.94) [M1W4]

(c) $\frac{d^2y}{dx^2} = 12x + 6$

When $x = 1.30$, $\frac{d^2y}{dx^2} = 21.6$ so (1.30, −5.94) is a minimum turning point.

When $x = -2.30$, $\frac{d^2y}{dx^2} = -21.6$, so (−2.30, 40.94) is a maximum turning point. [M1W1]

(d)

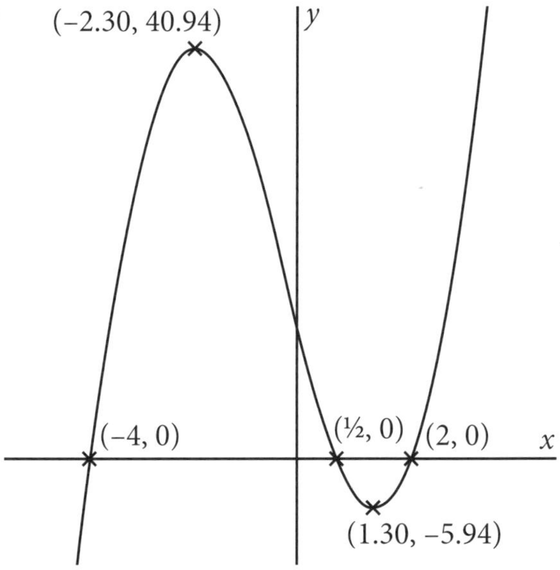

[1] for points, [1] for curve

7. $\int_0^2 (ax^3 - 2x + 6)dx = \left[\frac{1}{4}ax^4 - x^2 + 6x\right]_0^2$

$= (4a - 4 + 12) - (0) = 4a + 8 = 24$

so $4a = 16$, so $a = 4$ [M1W4]

8. (a) $D = x(x + 1) + (2 - x)(6 - x) = 2x^2 - 7x + 12$ [M1W1]

(b) $\frac{dD}{dx} = 4x - 7$, so $x = \frac{7}{4}$

$\frac{d^2P}{dx^2} = 4$ so it is a minimum turning point. [M1W4]

(c) 5 $\frac{7}{8}$ [1]